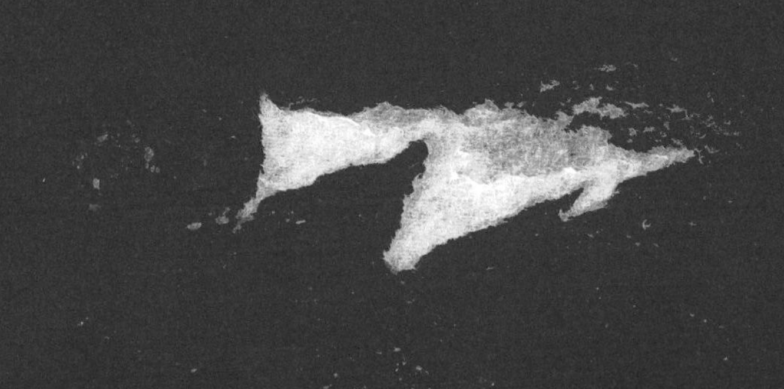

FOXES

Sally Morgan

W
FRANKLIN WATTS
LONDON • SYDNEY

First published in 2005
by Franklin Watts
96 Leonard Street
London EC2A 4XD

Franklin Watts Australia
45-51 Huntley Street
Alexandria NSW 2015

Produced for Franklin Watts by
White-Thomson Publishing Ltd
210 High Street
Lewes BN7 2NH

Editor: Rachel Minay
Designed by: Tinstar Design Ltd
Picture research: Morgan Interactive Ltd
Consultant: Frank Blackburn
Printed in: China

British Library Cataloguing in Publication Data
A CIP catalogue record for this book is available from the British Library.

ISBN: 0 7496 6066 X

Acknowledgements
The publishers would like to thank the following for permission to reproduce these photographs:

Ecoscene
FC, 1, 4–5 (Robert Pickett), 7 (Angela Hampton), 9 (Nigel Downer), 10 (Robert Pickett), 11 (Dennis Johnson), 13 (Ian Beames), 14 (Robert Pickett), 15 (Nigel Downer), 17 (Robert Nichol), 18 (Dennis Johnson), 19 (Robert Pickett), 22, 24, 26 (Robin Redfern), 28 (Robert Pickett);

Nature Picture Library
6 (Andrew Cooper), 8 (T Rich), 12 (Laurent Geslin), 16 (T Rich), 20 (Dietmar Nill), 21 (Laurent Geslin), 23 (T Rich), 25, 27 (Laurent Geslin), 29 (T Rich).

Contents

The fox

If you've been driven around a town at night, you might have glimpsed a dog-like animal crossing the road or jumping a wall. This is the red fox.

In the past, foxes were found mostly in woodland and open countryside. Today this animal is a familiar sight in British towns and cities. It can be seen roaming streets, parks and gardens, mainly at night.

ANIMAL **FACTS**

- *The scientific name of the red fox is Vulpes vulpes. Vulpes is the Latin word for 'fox'.*
- *The red fox is named after its reddish-brown fur. However, not all red foxes are this colour. Some have silvery-grey fur and others are almost black.*
- *The fox is related to domestic dogs, wolves and coyotes.*

Foxes are mammals

The fox belongs to a group of animals called mammals. Most mammals give birth to live young, which are fed on milk from the female. Male foxes are called dogs and female foxes are called vixens. The vixen gives birth to young called cubs.

VITAL **STATISTICS**

- *The body of an adult fox can be up to 90 cm long.*
- *Its tail is up to 50 cm long.*
- *It weighs as much as 10 kg.*
- *It lives for two to three years in the wild, but up to 12 years in captivity.*

bushy tail, called the brush

Dog foxes and vixens look very similar. Both have a bushy tail called a brush. It often ends in a white tip.

Reproduction

Foxes breed once a year. The dog fox and vixen mate in January or February.

The vixen is pregnant for about two months. Before her cubs are born, she finds and enlarges a hole in which to make her nest. The hole is called an earth. Her nest is in the deepest part of the earth, where it is warm and dry. She gives birth to four or five cubs in March or April.

The cubs stay close to the vixen in the earth so that they stay warm.

Tiny cubs

The newborn cubs are deaf, blind and completely helpless. The vixen stays by their side for the first two weeks, feeding them on her milk and keeping them warm. At two weeks the cubs' eyes open. The vixen doesn't leave the earth and relies on the dog fox to bring her food.

ANIMAL **FACTS**

- ***The mating call of the vixen is an eerie scream. It has caused some people – thinking somebody is being attacked – to call the police.***

These young cubs are about four weeks old.

Growing up

Soon the cubs are ready to leave the earth and explore. As they get older, they spend more time outside.

Fox cubs spend much of the day playing. Play helps them learn how to fight. When they get tired, they fall asleep in a pile. By the end of May the cubs are ready to go on hunting trips with the vixen. She teaches them what to hunt and how to catch it. The cubs grow quickly through the summer.

ANIMAL **FACTS**

- ***About 500,000 fox cubs are born each year but more than half die during their first year of life. Some are killed on the roads while others die from disease or starvation.***

Cubs learn to jump on each other. This is how they will catch mice and voles when they are older.

These two young cubs accompany their mother as she searches for food.

Leaving home

By October the cubs are ready to leave their parents and live on their own. They have to move away to find a place to live. Now the cubs have to find their own food, which can be difficult during the winter months.

Living in the countryside

Foxes are found throughout Britain, from the northernmost parts of Scotland to the southern tip of Cornwall. They are even found on some of the larger Scottish islands.

Foxes live in many different types of habitat, but mostly woodland and farmland. These habitats have plenty of food and shelter for them. Foxes also live in mountainous areas of Scotland and Wales.

Many foxes live in woodlands.

Hedgerows and overgrown areas on farmland give the fox plenty of cover to move around without being seen.

Foxes live on the coast

Some foxes live on sand dunes and muddy estuaries along the coast. The conditions along the coast are very different to those in woodland. These foxes have had to adapt to the different environment. They feed on animals that they find on the shore.

ANIMAL **FACTS**

- ***The foxes that live near the coast search beaches for crabs, dead fish and seabirds.***

Living in cities

Foxes are common in urban habitats such as towns and cities, where they are seen in streets, gardens and parks. They have got used to living close to people.

Many urban foxes do not need to hunt for food. They have learnt that people drop food on the street or throw it away in bins. Scientists have discovered that the urban fox is developing a different jaw to that of the rural fox as it scavenges for food rather than kills it.

Urban foxes investigate everything. This fox has found something interesting behind a drainpipe.

SCIENCE LINKS

How does a woodland habitat differ from that of a city? Where does the fox find food? How does the fox get around the city?

Foxes come into gardens

Most foxes are active at night but often urban foxes are seen on the streets during the day. Some people encourage foxes to visit their garden by putting food out for them. Many foxes have earths under garden sheds and by railway lines.

ANIMAL **FACTS**

- ***There are about 260,000 foxes in Britain, of which about 33,000 are thought to live in towns and cities.***

This urban fox is searching for food in a waste bin.

Foxes are omnivores

Foxes eat both plants and meat so they are called omnivores. (A carnivore, such as a wildcat or stoat, eats only meat.)

Foxes hunt and eat small animals such as beetles, as well as rabbits, mice, voles and even birds. On warm wet nights they hunt earthworms. They spend much of the day scavenging for the remains of dead animals. Foxes love fruits, especially blackberries, which are ripe in autumn. They can eat so many that their droppings become purple in colour.

The teeth of the fox can crush the bones of small mammals and birds.

SCIENCE LINKS

Have a look at your teeth in a mirror. Do you have long pointed canines like the fox? Have you got more or fewer teeth than a fox?

Foxes have 42 teeth. The long canines are much larger than the small incisors at the front of the jaw.

Teeth of foxes

The teeth of the fox are very similar to those of a dog. At the front are incisors that are used to hold food and pick meat off bones. There are four long pointed teeth called canines, which foxes use to grip their prey. Behind the canines are large jagged teeth that are used to crush bones and slice through meat. These are called premolars and molars.

ANIMAL FACTS

- ***A fox needs to eat up to 1 kg of food each day. Any food it can't eat is buried for future use. These 'larders' or food stores are used when the fox cannot find any food.***

Food chains

Foxes are predators. This means that they hunt and kill other animals. Each fox needs to eat many animals to survive.

Foxes hunt mostly herbivores such as rabbits. Herbivores are animals that eat plant foods such as leaves, fruits and roots. In general, the fox itself is not preyed upon by other animals so it is usually at the top of the food chain.

This vixen is bringing back food for her cubs.

ANIMAL **FACTS**

- ***Foxes learn to do a 'mouse pounce'. Small mammals such as mice and voles leap into the air when threatened. Foxes leap into the air and come down on them from above so their prey can't escape.***

Foxes and people

Foxes are killed by people. Some are shot by farmers because they kill farm animals, especially chickens and lambs. However many more foxes are killed on the roads at night. Also foxes suffer from a horrible skin disease called mange. Thousands of foxes die each year from mange, particularly in cities.

Each year more than 100,000 foxes are run over by cars and left to die by the side of the road.

Fox senses

Foxes have excellent senses, which they use to find food.

Foxes are mostly nocturnal, which means they are active at night. Their senses help them to find their way around in the dark. Foxes have good night vision and can see better in the dark than people can. Their eyes are large and forward-facing, which allows them to judge distances when they are hunting. They have large ear flaps to funnel sounds into their ears. They can hear the slightest rustle in the undergrowth caused by mice and voles.

The fox uses all its senses to find prey.

This fox has found an interesting smell on the ground.

Smell and touch

The fox has a long muzzle and a large nose. This helps the fox to detect smells on the ground and on walls and plants. It has stiff hairs around its muzzle called whiskers, which are very sensitive to touch.

ANIMAL **FACTS**

- ***The hearing of a fox is so good that on a warm night a fox walking slowly across a field can hear earthworms moving through the grass.***

science **LINKS**

How good is your night vision? Stand in a well-lit room at night and then turn the light off. How long does it take your eyes to get used to being in the dark? After a while you should be able to see far more detail than you did at first.

Movement

Foxes have four legs which end in feet. Each foot has four toes with long claws.

The fox has an internal skeleton which is surrounded by hundreds of muscles. The muscles are attached to the bones and when they contract (get shorter) they pull on the bone, causing it to move.

Foxes trot at speeds of between 6 and 13 km/h and can reach speeds of 48 km/h when running at full speed.

science LINKS

See how muscles work in your body. Bend and stretch your arm. Feel how the biceps muscle in your upper arm changes shape. The biceps muscle is attached to a bone near your shoulder and to a bone in the lower arm. To bend your arm, the muscle contracts and becomes short and fat, pulling the lower arm up.

Moving around

Foxes can walk, run, jump, climb and even swim. They have lightweight bones, which means that they can move more quickly than other mammals of a similar size. When it runs, a fox holds its tail out behind it. This helps it to balance. Foxes are agile animals and they can leap up to 2 m in height.

Foxes are very agile animals and they can jump and spin around in a fight.

Fox homes

Foxes live in a home called an earth. This is a large burrow or hole in the ground.

Foxes do not use the same earth all the time. They have several earths, which they use at different times of the year. An earth may have several entrances and more than one chamber or room underground. There is usually an entrance chamber, another for storing food and the deep chamber where the vixen makes her nest.

animal CLUES

Look out for a fox earth. It is a large hole in the ground, often under a shed or root of a tree. A pile of freshly dug soil, feathers and bones are often seen near the entrance.

These cubs are sitting beside the entrance to an earth.

Finding new earths

Sometimes a fox digs out a new earth, digging the soil with its sharp claws. At other times the fox looks for holes left by other animals such as rabbits or badgers.

For much of the year foxes do not use their earths, preferring to sleep curled up under a hedgerow or in a sunny spot.

Territories

A fox lives in a certain area that is called its territory. Foxes do not often go outside their territories.

A territory is occupied by one dog fox and one or two vixens and their cubs. They live in the territory most of their lives and find all their food within its boundaries. Their 'larders' are scattered all over the territory. At the end of summer the parent foxes chase their fully grown cubs away. The cubs have to find their own territories.

Well-worn paths link all the foxes' earths in a territory.

These two urban foxes are standing at the point where their territories meet. Eventually one will chase the other away.

Territories vary in size

In places where there is plenty of food, the territory is usually small. In other places, where food is much scarcer, the territory is much larger. Most territories are between 2 and 6 sq km in size. Urban foxes tend to have smaller territories than rural foxes.

ANIMAL **FACTS**

- ***In the Scottish Highlands, because of the lack of food, territories can be as large as 40 sq km.***

Fox communication

Foxes communicate with one another using smell and sound.

Smell is important in marking the boundaries of a territory. Foxes have a very smelly urine, which other foxes recognise. A fox also has scent glands on its tail and between its toes, which release a smelly substance.

ANIMAL **FACTS**

- ***The fox wags its bushy tail, just like a dog. This is thought to help spread its scent around.***

This vixen is marking the ground with her urine.

Foxes can make a lot of noise such as screams and yowls when they meet each other.

Foxes have many calls

Foxes have more than 20 different types of call, which allow them to keep track of each other as they roam around. Each call has a meaning. Vixens have a hoarse call and they may wail. The dog fox has a short clear yap which it repeats three times. Foxes growl when they are not happy and make a cackle when they are angry or frightened. Foxes are particularly noisy during the mating season, when they scream at night.

science LINKS

Pets such as dogs and cats communicate with us and with each other. Watch a dog or cat to see how it communicates. Compare this to a fox.

Fox stories

Many stories have been written about the fox and how it tricks other animals.

Hundreds of years ago, Aesop wrote this fable (story) about the fox and the goat.

The fox and the goat

A fox fell into a deep well and couldn't climb out. A passing goat asked the fox what he was doing down there. The fox said, 'Haven't you heard? There's going to be a drought, so I've come down here to be close to water.' The goat thought it was a good idea and jumped down too. The fox immediately jumped on the goat's back and escaped from the well. 'Goodbye,' said the fox, 'be more careful next time.'

There is a Scottish tale of how the fox rids himself of fleas.

The fox and the fleas

The fox was covered in fleas and he was desperate to get rid of them. So he took a mouthful of wool and waded into the river. As he got deeper, the fleas moved to his head to escape the water. Finally they ran across his nose and onto the wool. The fox let go of the wool, which floated downstream with all the fleas.

Fox facts

Fox family

The red fox belongs to a group of mammals called Carnivora. This group includes bears, wolves, big cats, stoats and otters. The red fox is the most widespread and abundant member of this group in the world.

MAIN FEATURES OF THE FOX

- ***The fox is a mammal.***
- ***The male fox is called a dog and the female fox is a vixen.***
- ***The fox lives in the countryside and in towns and cities.***
- ***Its home is called an earth.***
- ***The fox is an omnivore.***
- ***The vixen gives birth to four or five cubs in spring.***

Did you know?

There are 21 species or types of fox, including the Arctic fox, fennec fox and bat-eared fox.

Fox websites

UK Safari

www.uksafari.com/redfox.htm

Website encouraging young people to learn about British wildlife.

Derbyshire Fox rescue

www.derbyfoxes.org

Volunteers who look after sick and injured foxes. An informative site with good links.

Wildlife Online

www.wildlifeonline.me.uk/red_fox.html

Page packed with information about the red fox worldwide.

Glossary

adapt get used to

agile able to move quickly and easily

carnivore an animal that eats only meat

coyote a North American wild dog

cub a young fox

dog fox a male fox

earth the name given to the home of the fox

food chain feeding relationships between different organisms, for example plants are eaten by rabbits and rabbits are eaten by foxes

gland a group of cells that produce a liquid, similar to sweat glands in the skin

habitat the place where an animal lives

mammal an animal that usually gives birth to live young. The female mammal produces milk for her young

mange a disease that can kill the fox, which is caused by microscopic parasites called mites that invade the skin

mate reproduce

muzzle the part of an animal's face that includes the nose and mouth

nocturnal animals that are nocturnal are active at night and rest during the day

omnivore an animal that eats a mixed diet of plants and meat

predator an animal that hunts other animals

pregnant a female animal is pregnant when she has a baby or babies developing inside her

prey an animal that is hunted by other animals

scavenge feed on dead and decaying food and food in rubbish

territory the range or area of land in which an animal lives

urine the water passed out of the body of an animal

vixen a female fox

Index